KEEPING UP with the EGYPTIANS

KEEPING UP with the EGYPTIANS

Building Pyramids Revised

James A. Kizer

Main Illustrator: William Perry
Raft Illustrator: Wallace F. Duncan

Library of Congress Control Number:		2017904602
ISBN:	Hardcover	978-1-5434-1096-9
	Softcover	978-1-5434-1097-6
	eBook	978-1-5434-1098-3

Print information available on the last page.

Rev. date: 04/18/2019

To order additional copies of this book, contact:
Xlibris
1-888-795-4274
www.Xlibris.com
Orders@Xlibris.com
703823

Contents

A PROBLEM UNSOLVABLE

There's a great story told about mathematical scientist George Dantzig.

During his first year of graduate school at Berkeley, he arrived late to class and saw two math problems on the board, which he assumed to be homework. He copied them and began to work on them.

Finding them to be harder than usual, he struggled for several days to complete them. Finally, he finished and turned them in, thinking he had finished the assignment late.

A few weeks later, his professor paid him a visit, explaining to him that the two problems written on the board were not an assignment, but an example he had given the class of two famous unsolved problems in statistics.

Dantzig had solved an unsolvable problem. His solution was soon published in a mathematical journal. It was the beginning of a great career for this mathematician, who would go on to become the father of linear programming.

What if Dantzig had arrived on time to class that day? What if he knew upfront that the problems were unsolvable? Would he have tried to solve them?

—Author unknown

FOREWORD

The Egyptian pyramids may have been a lot easier to build than many believe. This theory gets down to the "nuts and bolts" and avoids religious and spiritual meanings and doesn't try to answer methods used to mill stone to perfection, but do note if they can mill stone, then they can mill wood to meet nearly any need too. I found common Egyptian hieroglyphs and applied them to my interpretation. This theory sticks to relatively simple means and mechanical functions and was developed within the mind of me, the author. Any coincidence to any other theory is just that--coincidence. I am merely putting forth an independently thought-up plan.

And to back up for one moment, I would like to say that as we learn more about the capabilities and ingenuity of the ancients, we have to come to the realization that these people are far from cavemen. They were highly intelligent and capable of great skill.

My hope is that you will gain a new perspective of Egyptian construction as a result of reading this small book.

Sincerely,
James A. Kizer

GETTING STARTED

In order to assemble the Egyptian pyramids in a simple, efficient operation, they gained an understanding of three basic principles. These principles may have been learned hundreds or thousands of years apart. But they converged at a heightened level in order to help build these great structures.

WATER

ONE PRINCIPLE

At least three things happen to a rock placed in water--it sinks, it gets wet, and most importantly, it gets lighter. Generally, all mass has buoyancy.

Buoyancy--the ability or tendency for matter to weigh less after being submerged in liquid.

By the numbers, a cubic foot of stone weighs about 170 lb. Water weighs in at about 62 lb per cubic foot. Submerging a stone into water lightens the stone equal to the weight of the water it displaces. Thus, a block of stone weighing 170 lb now weighs 108 lb. This makes a stone more than ⅓ lighter simply by putting it in water.

Let's also keep in mind the Egyptians obviously knew how to float something on top of the water, and by the numbers, a cubic foot of air has 62 lb of lift in water.

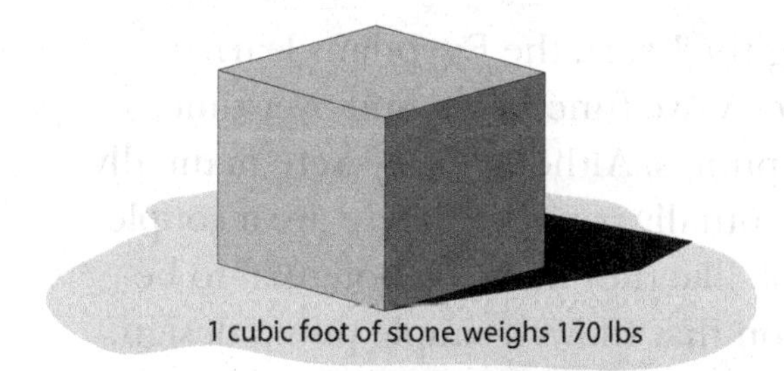

1 cubic foot of stone weighs 170 lbs

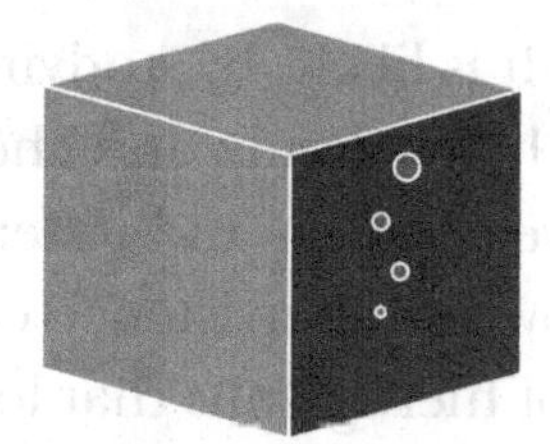

1 cubic foot of water weighs 62 lbs

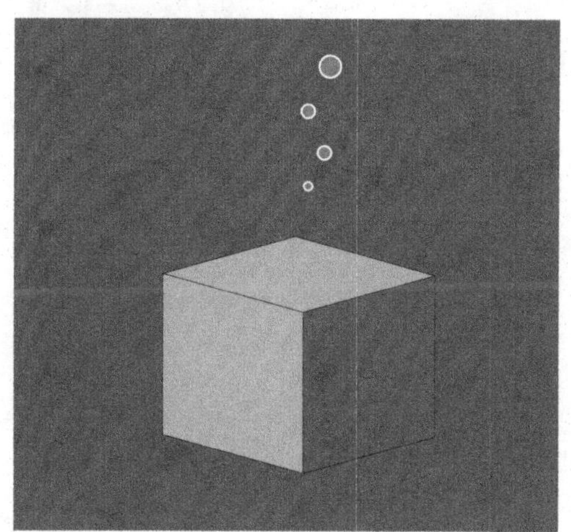

1 cubic foot of stone weighs 108 lbs in water

ANOTHER PRINCIPLE

It is likely, by studying the heart, the Egyptians learned how a pump and check valve functioned and over time, developed air and water pumps. Although they were manually powered, persistence eventually payed off. Here are a couple of hieroglyphs that look like they have the potential to be pumps and an illustration that shows a simple pump design.

AIR PUMP

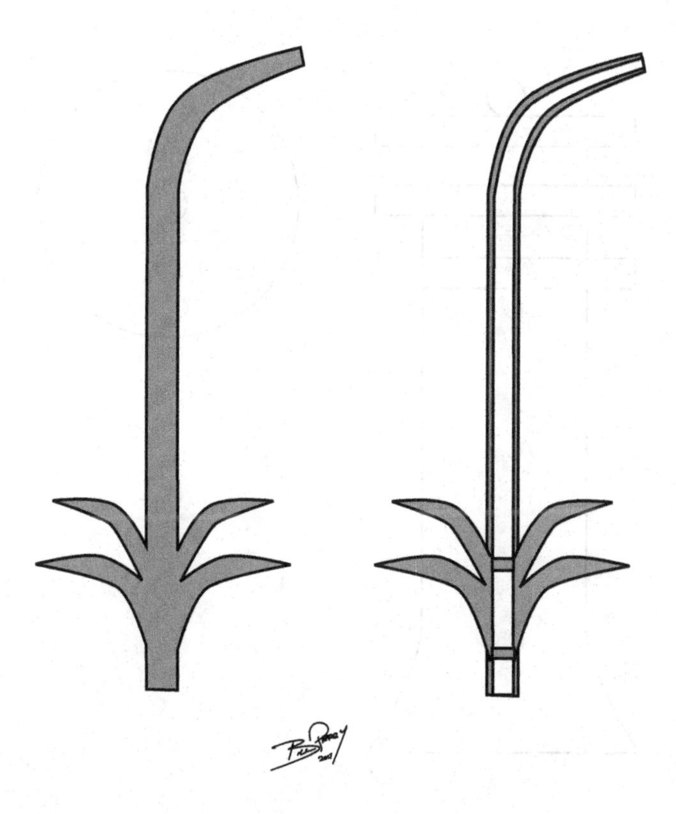

This hand pump had bellows on one end and the hooked end directed air into a working position. Perhaps, first used to stoke a fire.

PUMP

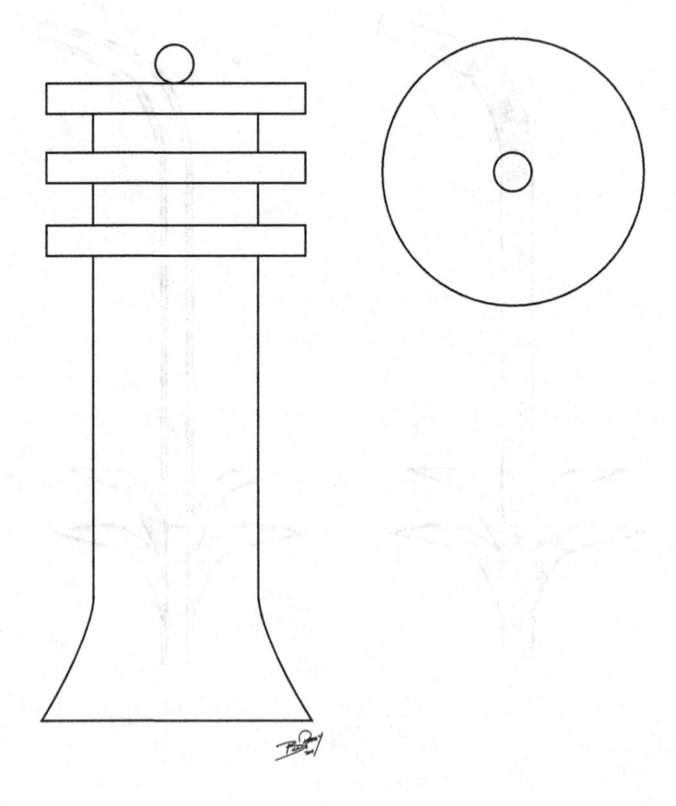

This pump telescoped in and out and worked with air or water.

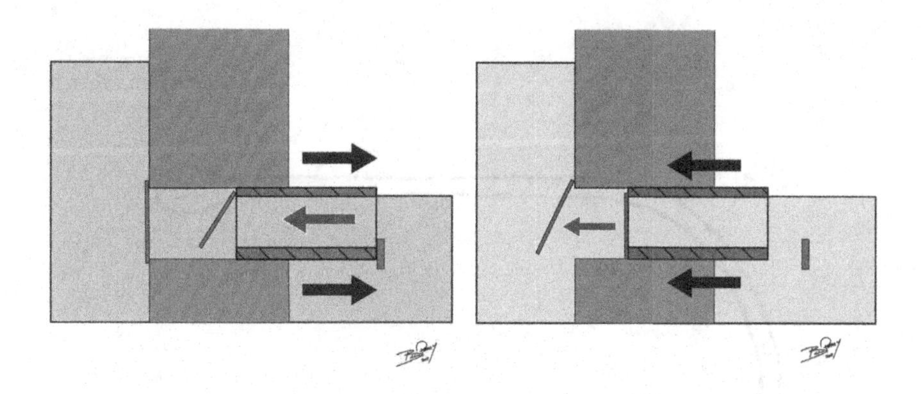

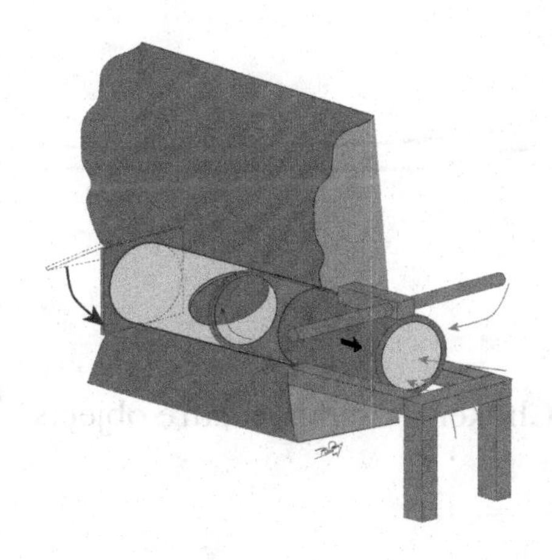

ROPE CHOKER

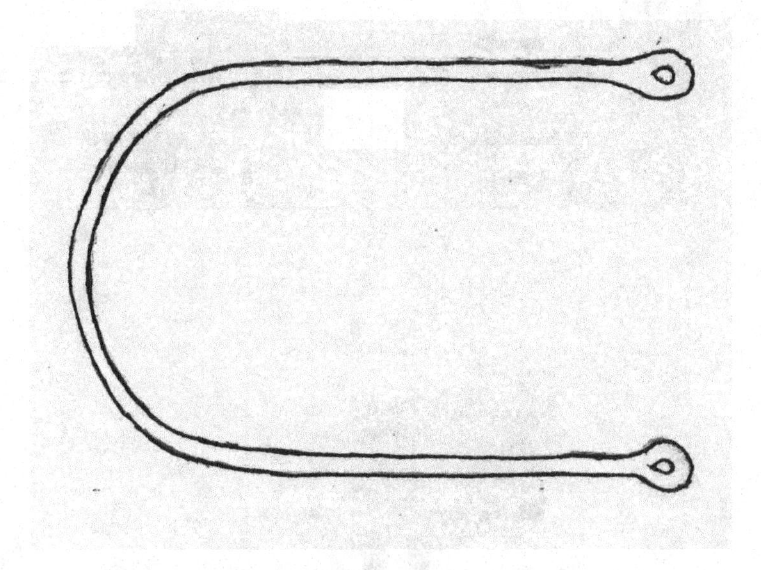

Choker––used to secure objects.

RAFT FRAME

Raft frame—the backbone of the raft.
It tied everything together.

FLOAT BOXES

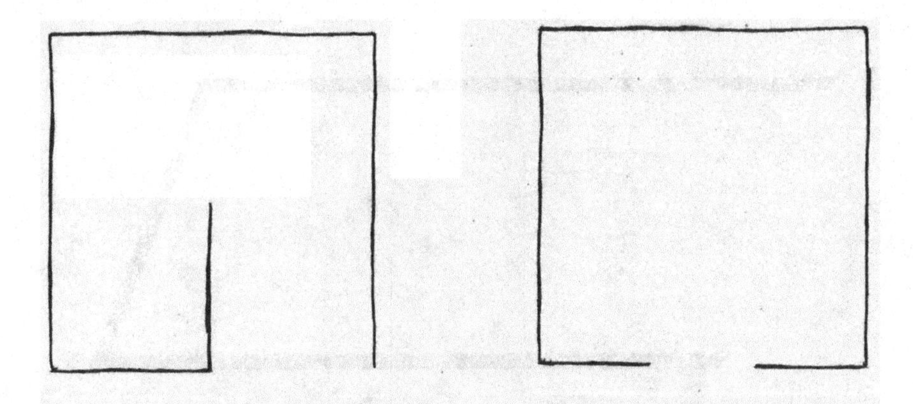

The float boxes were well made of timber and lined on the inside with tar and animal hide. They were used in conjunction with a raft frame and chokers to float heavy payloads. The spiral box was used for inclined lifting and roll maneuvers, while the middle open box was used for standard transport. Some people believe these symbols were for a house or a dwelling. The workers may have slept in them after a long day on the job. One 3' x 3' x 5' float box had a lift capacity of 2,790 lb.

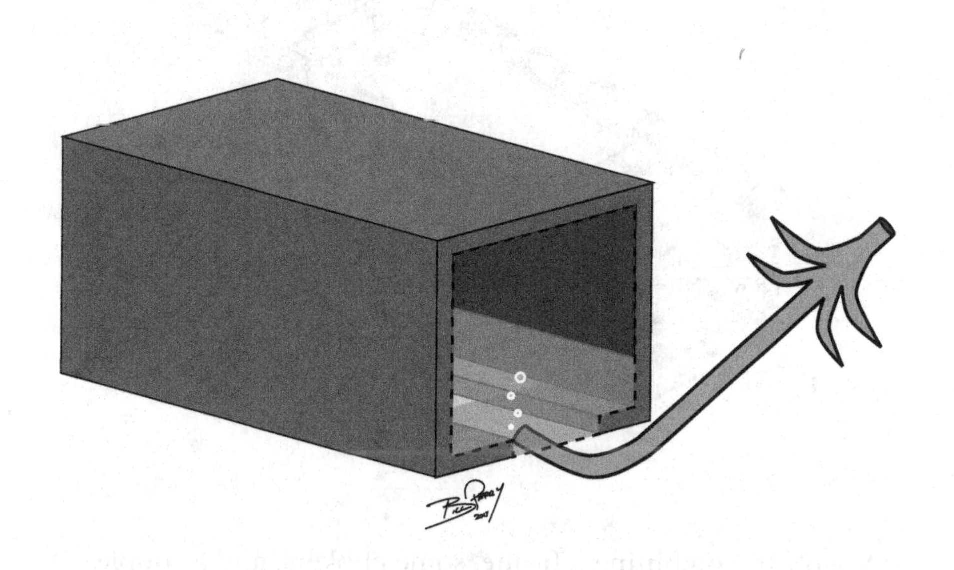

THE RAFT

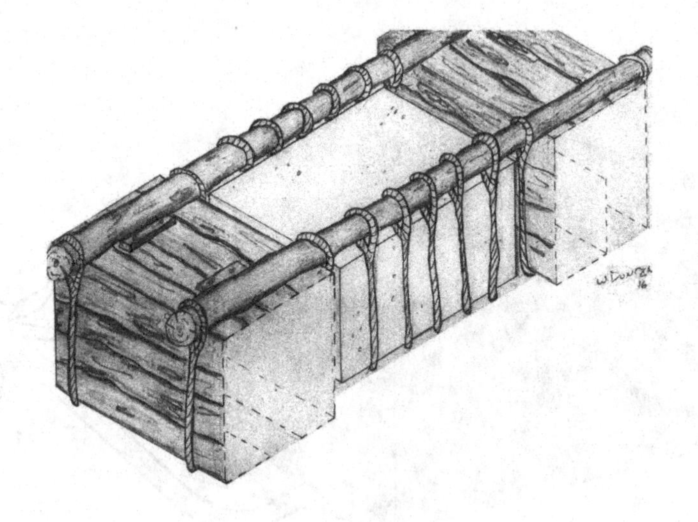

Simply by combining a frame, some chokers, and a couple of float boxes, we have the ability to haul an 8,000lb payload with little effort. The raft with payload takes up an area 6' x 13' and requires about 4' of water.

This may also be the system they used to ship casing stone from over 100 miles away. An up scaled version of this system could have been used to transport obelisks and other big loads, from the stone yard to the worksite. Floating submerged stone would have been the key of success in almost all above-ground Egyptian construction. A large pyramid jobsite could require 200 rafts to run efficiently.

BIG LOAD

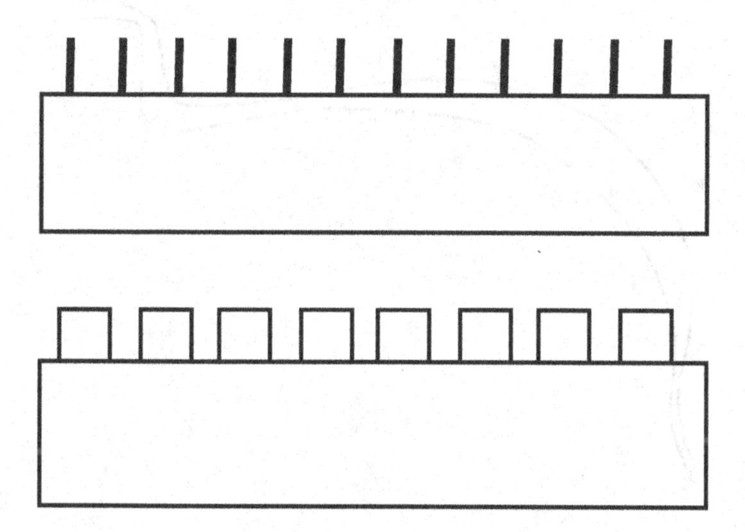

This hieroglyph seems to depict a side view of a large stone with the ends of multiple frame cross members exposed on the top.

WATER TRANSPORT

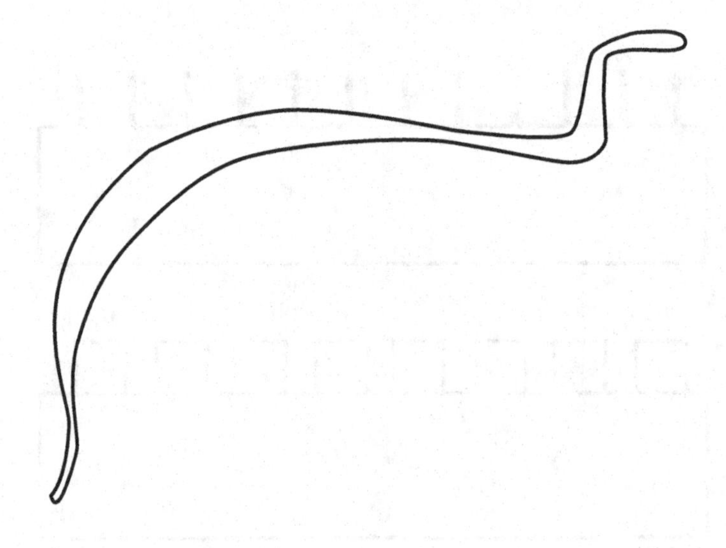

The snake in the water hieroglyph, may have originally been designated to mean shipping several rafts tied together. They didn't float down the river in a straight line, they bowed and bent and looked like a large "snake" in the water. Perhaps, it later became the common term for shipping by water.

LAST MAIN PRINCIPLE

The Egyptians also learned to build watertight. One example––their boats. Some had been estimated to haul 40-ton payloads. Made of wood, they had to be built for the task. Another example––the float boxes, designed to be durable. A leaky box would have burned valuable time and energy. Lastly, they learned how to make stonewalls and perimeters liquid tight, which allowed them to pool and store water.

PREPARING A JOBSITE

All big builders know that the better you set yourself up on the jobsite and with your suppliers, the better the job should go. At least seven things had to be addressed before the main construction can begin.

First of all, food and housing for 3-5,000 workers had to be steadily provided or supplied.

Second, there had to be a dependable water source to the pyramid––whether it was brought in by trench, elevated canal, tunnel, well or trapped during flood season. Egyptians needed a sound water supply to make their jobsite flow. The hieroglyph for water was one of the most used in all Egyptian writing. Why all the concern about water––unless it was being used extensively?

Third, an exact location of the structure needed to be marked out. So the fourth thing, a watertight perimeter wall, can be built around the foundation.

Other things that had to be done, a port area had to be in place to receive incoming supplies. A good example of this may be the structures found near the old Nile riverbed in front of the Sphinx. They also needed a sound pathway to haul materials to the worksite, a.k.a. the causeway. It was built straight, as if on a line, The Egyptians didn't mess around. It even self-adjusted to the Nile like a long boat ramp during flood season. And last, they had to have sound sources of stone, lumber, mortar agents, tools, skilled manpower, and funding to complete the project.

SKILLED CRAFTSMEN

Every day, tradesmen such as carpenters, bricklayers, boilermakers and electricians go to work on projects big and small. It was no different back then. They had stonecutters, blacksmiths, carpenters, sculptors and laborers, to name a few. Keeping their skills from being common knowledge, helped insure steady employment, in much the same way as today. One other benefit that comes from doing the "big move" or "big job", is the "star" status (admiration) from others that get to be witness to these events. Even to the point of setting trends or styles amongst the "regular" folks, construction jargon is always cool and can seem larger than life.

STAFF

The angled end of the staff was used to pry and control loads. Even modern day push/pull rigging sticks have adopted this shape. The U shape on the other end could have wiggled wedges out, controlled loads in the water or staved off an occasional crocodile while out on the Nile. Clearly, it could have been used for self-protection.

WEDGE

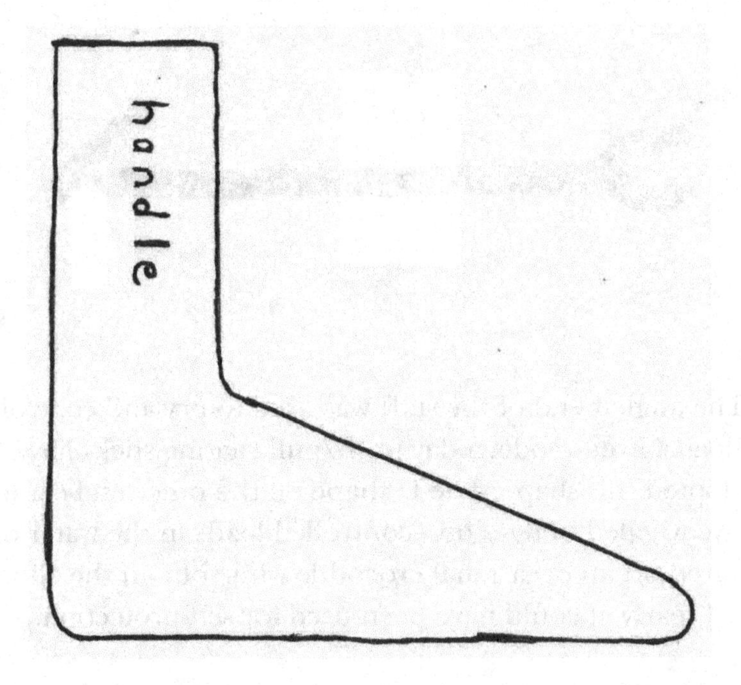

The wedge had an extended handle to keep the worker's face out of the water while positioning it under the load. The blunt end helped the wedge pop out while removing it, a sharp end tends to get pinched. The wedge may have looked like a foot to remind the workers to watch their toes!

MUD GUN

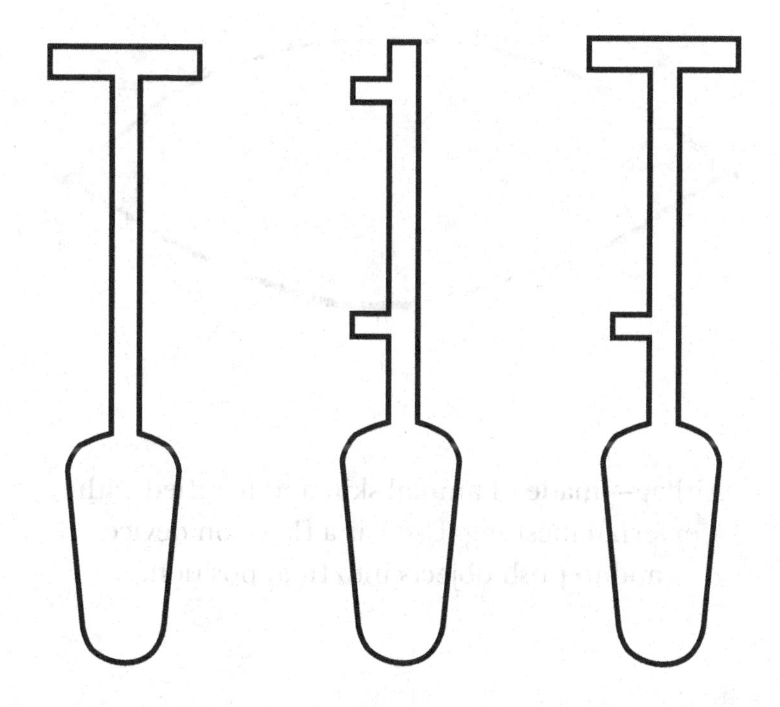

In much the same way as a caulk gun works today, this device shoved large amounts of mud deep into crevices where mortar or sealing agent needed to be applied. It could have delivered 1 to 2 gallons of product in one push.

AIRBAG

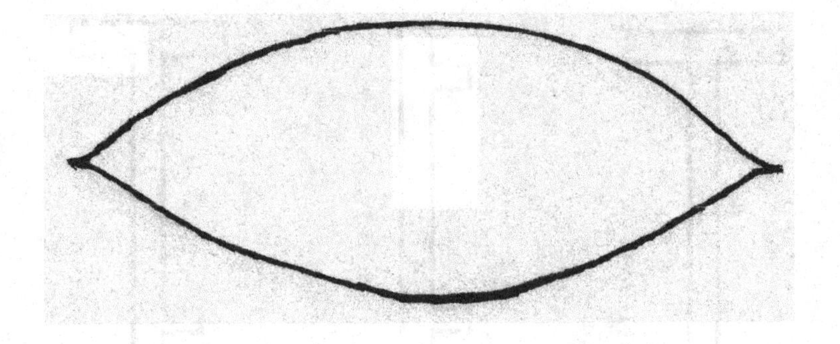

Airbag––made of animal skin and fortified with external meshing. Used as a flotation device and to push objects into final position.

There are so many potential choices when it comes to hieroglyphs that look like tools. All the hieroglyphs shown in this book that are "air filled", were depicted on the rear wall of Ramses 1st burial chamber painted white on the inside of them. Generally, background color was blue/ grey. Could this be a clue? It is reasonable to believe, that hieroglyphs started out as common items used regularly, and some, over time, became more symbolic to spiritual meaning when that item became obsolete or immortalized.

Two more interesting "air filled" hieroglyphs will be revealed later.

[illegible]
[illegible] look like [illegible]
[illegible]
[illegible]
[illegible] Generally, [illegible]
[illegible] it is [illegible] to believe that
[illegible]
[illegible] because [illegible]
[illegible]

[illegible]
[illegible]

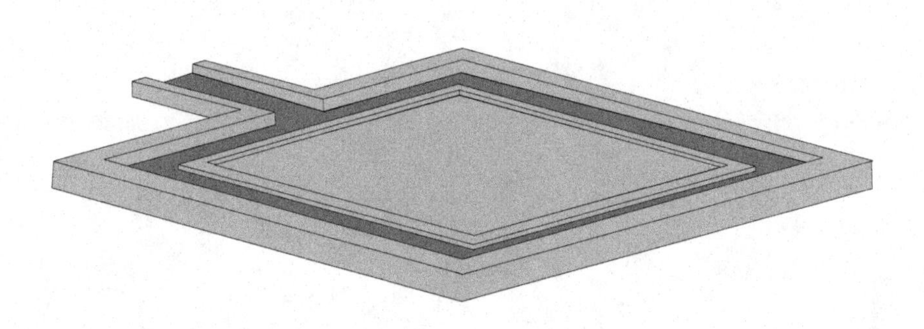

THE FOUNDATION

Once the foundation was marked out and the perimeter walls were sealed tight, water was pumped into the makeshift mote around the foundation in order to establish the starting elevation. Then, stone was chiseled to a near-perfect shelf-like plateau ledge all the way around the foundation's perimeter. Any low spots would have been fitted with large stable slabs of stone to create one continuous ledge.

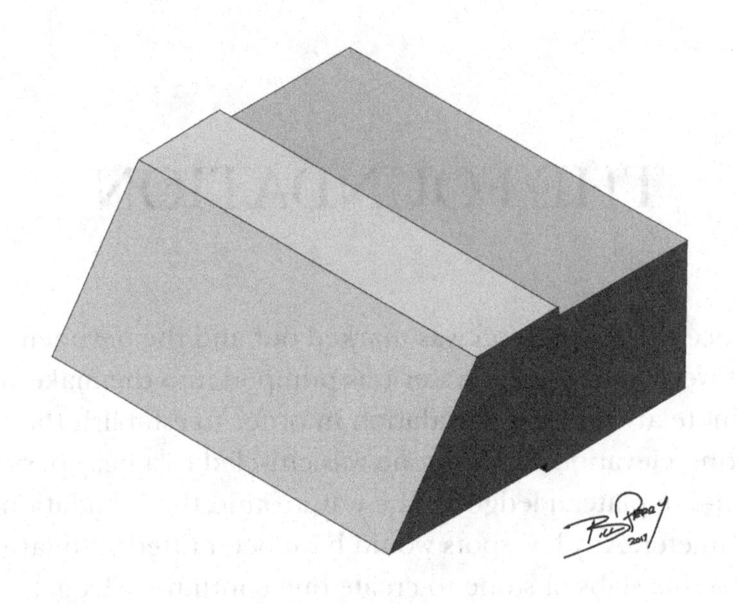

CASING STONES

It seems, the casing stones were premade off-site and shipped to the worksite ready to be placed. Here are some interesting details noticed about the existing casing stones on Kafre's pyramid at the Giza Plateau: The triangle part of the stone was only a small part of the actual stone. The more substantial portion of the stone points inward toward the core. Another detail observed, is how the casing stones stagger each other perfectly up all 4 corners. This also ensured that the middle casing starts out staggered. It's clear the corner stones were set first, then, the middle casing were filled in afterward. And possibly the biggest observation of all is the ridge cut into the top and (presumed) bottom. These ridges interlocked the casing, making them stable and created a great surface to seal casing stone to casing stone with mortar or sealing agents.

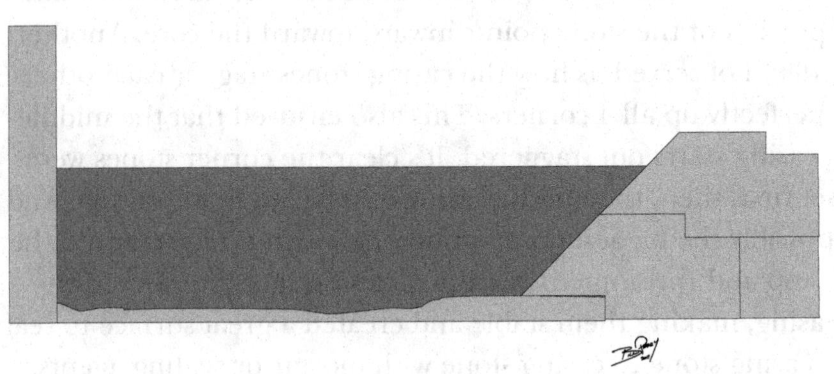

GROUND LEVEL

When the foundation surface was ready, more water was pumped in to continue filling the perimeter mote until the first two or three rows of casing stone are set and sealed. Also built at this time were four lift chambers. They were placed at the center of all four sides and attached to the set casing stone.

THE LIFT CHAMBER

The lift chamber is a small box like structure that rafted stone and other supplies are loaded into and released to float upward from, during the project. Picture the lift chamber, as a giant check valve. The lift chamber also serves to raise and lower water levels, by pumping in or releasing water.

FLOODING THE MIDDLE REGION

At some point after the first three rows of casing stone had the ability to hold water, the pyramid's middle region was flooded to allow work to continue upward. As the perimeter casing stone was being set, the common fill stones were brought in to fill the core, and the water shafts were extended.

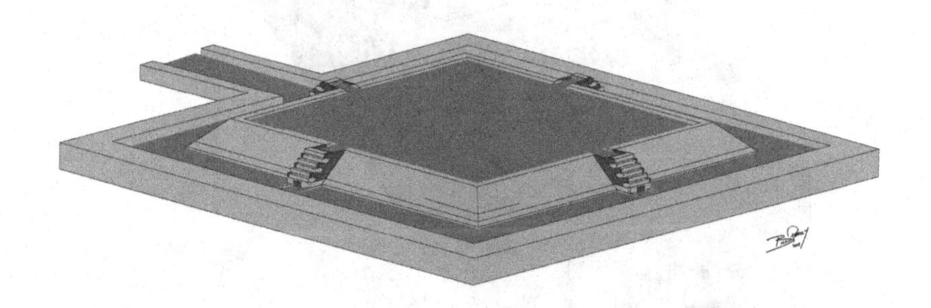

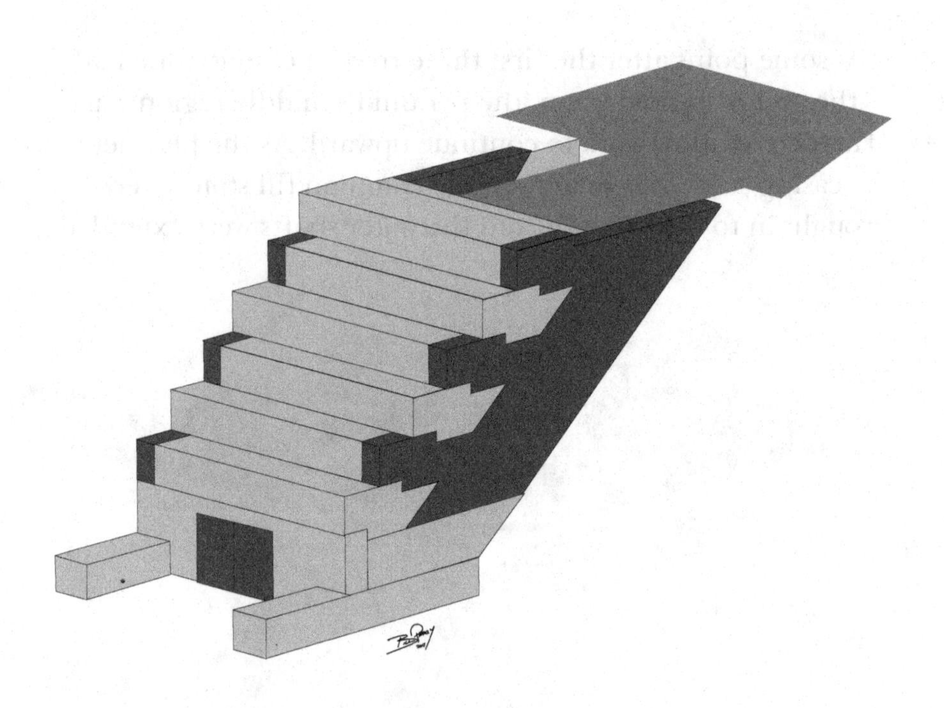

THE WATER SHAFT

The water shaft was an extension of the top of the lift chamber that was added to as the project rose upward. It was a popped-out area attached to the side of the pyramid. By taking two casing stones, sliding them out, flipping them over, and adding two sidewalls made of heavy timbers (16 to 20 feet long), three important things were accomplished. First, it made a clear route to elevate the rafted stone. Second, the casing stone's square backside now formed a stairway the workers can use when timber ties were added to secure the stone part of the shaft. And third, the casing stone's exterior face now created a smooth surface on the under topside of the water shaft for ascending objects to slide along without any hang-ups or snags.

TO REHASH AND MOVE FORWARD

A rafted stone was placed into the lift chamber. Front doors were closed. The chamber was pumped full of water until the upper doors open and the rafted stone ascended into the water shaft, allowing it to float upward, diagonally, until it reached the top surface of the water. Then, the rafted stone was retrieved and brought near its planned location. If it was a common center fill stone, it was slowly lowered to its location by deflating the float boxes and placing wood wedges under the stone. Next, the raft and rigging were pulled clear to allow the wedges to be wiggled out, thus setting the stone in its permanent location.

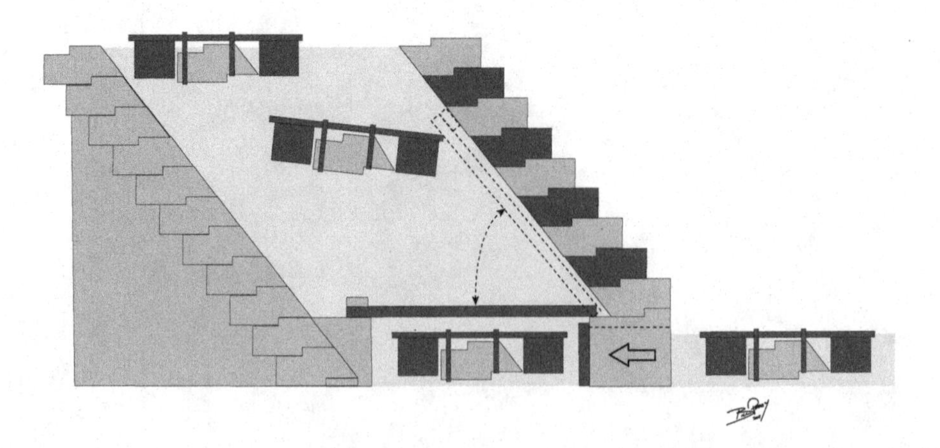

When a rafted casing stone reached the top of the project, it would have been worked a little different. It was brought near the area that it will be placed and set onto a giant counterweighted lever that was rafted into place, loaded, with the casing stone's point sticking up. Then, it was slowly lowered onto the cribbing or wedges so that rigging can be removed. Mortar or sealing agents were applied, and the piece was lowered by wiggling out the wedges. If needed, the casing stone was tapped (rammed) with a large piece of lumber to shift it into its exact final location.

And after the stone was removed from the raft, the empty raft was slid down the side of the pyramid, retrieved, and then refilled with another piece of stone.

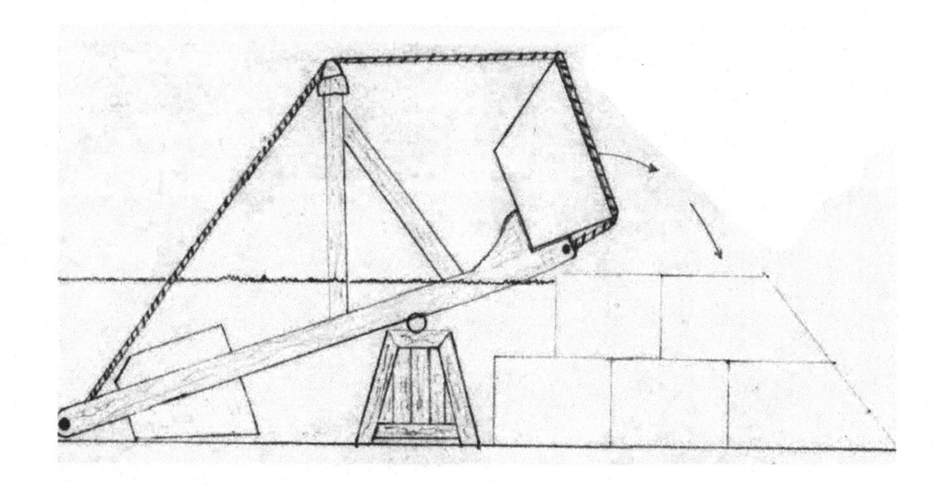

TOPPING OUT

As things got tighter at the top of the structure, it started to cause new problems. The queen's pyramids may have been built to help rehearse the top of one of the big pyramids. They may have also been a test site, for it was noted that they used two different building styles to build them, possibly trying to establish which building style was most dependable.

CAPSTONE

There must have been one heck of a celebration when the capstone was set. And how nice it was to be able to deliver the capstone to within a few feet of its final location without breaking a sweat!

THE JOB ISN'T DONE YET

At this point, all of the main upward lifting was complete, and it was time to disassemble the water shaft by discarding the timbers and fitting the casing into its final location. It was also possible that there may no longer be need for water pressure in the pyramid.

WATER PRESSURE

Water pressure concerns always enter into this conversation. According to water pressure theory, if unchecked at 460', a water column can produce about 200 psi. The Egyptians had a few simple ways to control water pressure. First of all, due to the converging shape, half way up vertically, 4/5's of the pyramid is set. This reduces the amount of water mass on top. By itself, it has little effect on lowering pressure. Combine it with a tightly packed core and water pressure plummets by more than 80%! Due to the challenges at the Bent Pyramid, other tactics became necessary. "Weep zones" (friction) could have been applied by packing a thick clay-like/mortar layer across one complete elevation of stone. This, would have slowed the water's ability to freely flow through the sand, rubble and stone core. A "lock" and transfer area in the water shaft would also be necessary at the same elevation. Good places to look for clues would be near the "bend" at the Bent Pyramid and at the bottom of the Grand Gallery in the Great Pyramid. If some "seep" water was directed into the lift chamber, it would have refilled the chamber for a quicker launch.

One other interesting detail that came out of the water pressure experiment...if the core is packed tightly, it only takes a small amount of water to fill it up. This means, the pyramid could be completely drained. Repairs or adjustments could be made. And then, it could be refilled. Perhaps, in only a few days. Lastly, water acted gravitationally more in a vertical manner. Thus, distributing weight of water and shaft directly to its location on the pyramid and not bearing down on the bottom of the shaft and lift chamber.

Water Column Pressure Chart

P.S.I.	Height Ft.
1	2.306
50	115.33
100	230.666
150	345.99
200	461.332

TAKING A LOOK AT THE THEORY

It is easy to see that moving stone in this manner was much easier. First, it required less people--especially less people working really hard. Second, it seems reasonable to believe that with four water shafts in operation, a stone reached the top of the project every five minutes. Third, every time a fill stone was placed, the water level came up!

Two more interesting hieroglyphs:

FLOATING POD

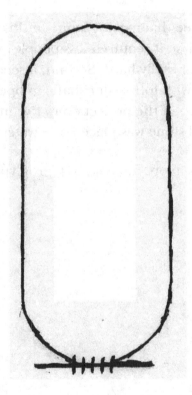

Floating pod--made of clay, like a giant vase or engineered with lumber, covered in animal skin, and tied at the bottom. Used to haul materials, tools, food, and people up the water shaft. Possibly the first submarine (nonmotorized).

AIRBAG FOR BREATHING

This symbol is seen being placed up to their mouth and may have been used in the longer rides up the water shaft. This hieroglyph was also shown with three ring-like objects stacked under the bubble on top of the T part. These rings could have been stone or metal weights used to help the airbag sink down to the work area when working under water. When I found out scholars believe this symbol to mean "life," I was pleasantly surprised!

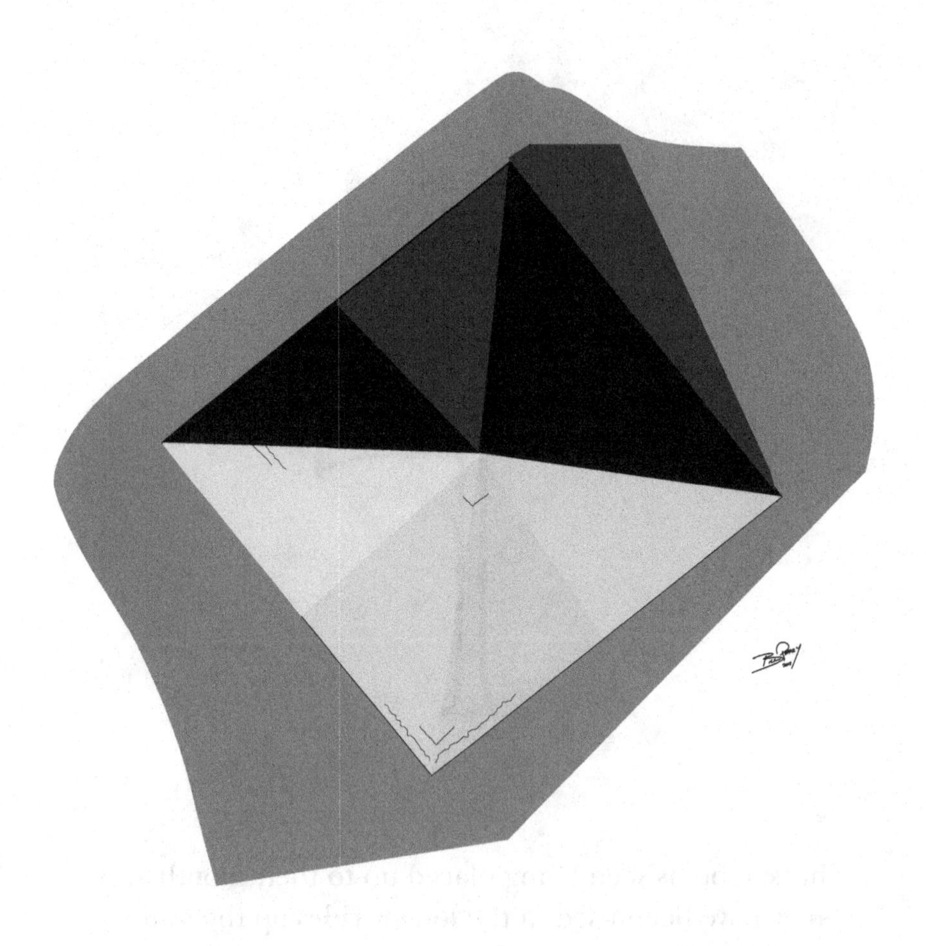

THIS THEORY APPLIED TO THE GREAT PYRAMID

Appling this theory to the Great Pyramid starts to answer or provide reasonable explanations to a few questions or observations. First of all, the Great Pyramid is actually considered to have eight sides. The four main sides actually concave in a little, forming eight. The water shaft could have left a mark like that because of two reasons: it was a work zone that was completed last, and there was an unrealized expansion of the shaft wall timbers after they were soaked in water. A 16-foot piece of lumber can grow up to 3 inches in length. Giving up 2 or 3 inches each elevation turned into a few feet (along with some major aggravation for the builders) before long, and that's how the vertical exterior grooves were formed on the Great Pyramid.

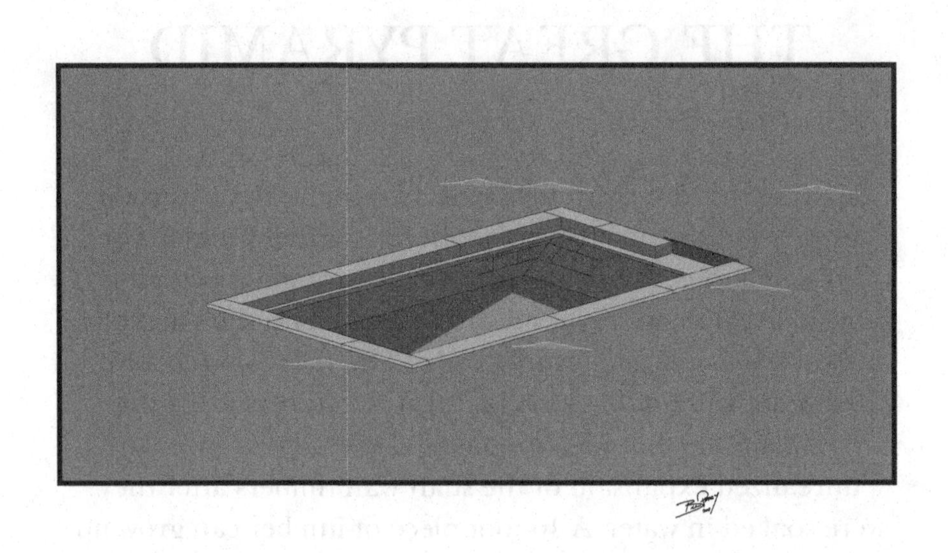

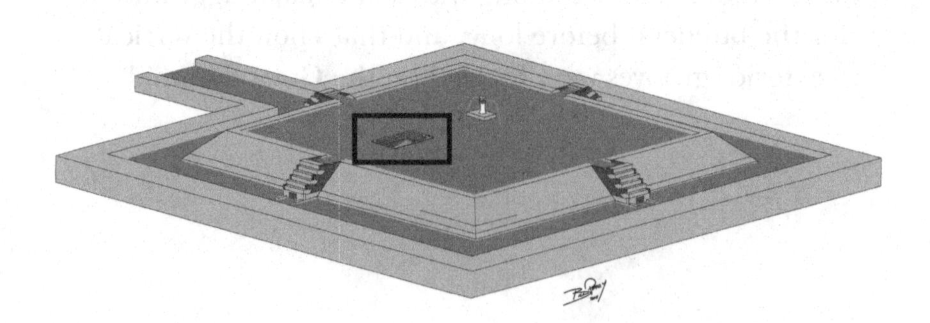

THE GRAND GALLERY SLABS AND AIR SHAFTS

Secondly, this theory helps explain what role the Grand Gallery slabs played, how they got so high in the structure, and what purpose the well shaft and air shafts served.

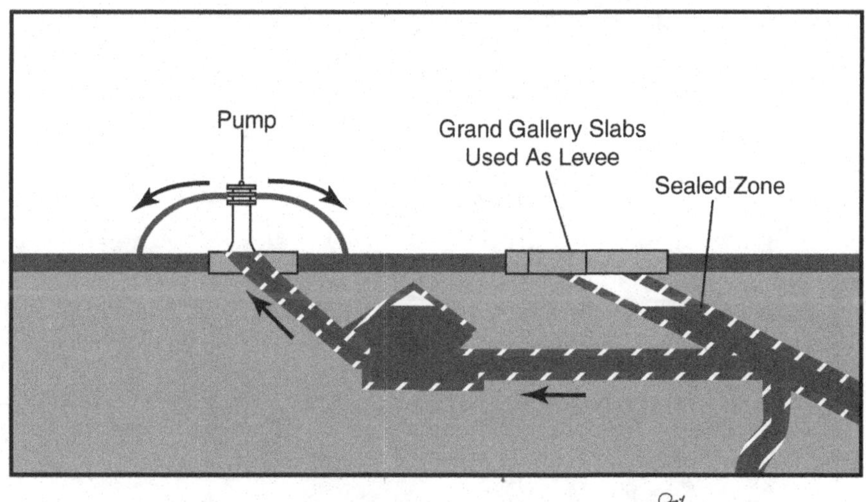

The Grand Gallery slabs were brought in early on when the first pieces of casing stone were being set. Their purpose was to create quick, stable, and temporary levees around the internal tunnel work zone. These internal work areas are built to be watertight even from the filler core areas of the pyramid. This was why the tunnels were built so precisely.

These areas took longer to complete, and control over water elevation in the tunnel work zone was necessary. This was why there was a well shaft or air shaft vertically placed within all of the internal tunnel construction from the near bottom all the way to the top of the King's Chamber, including the roof. The well shaft and air shafts were used to lower the water level in the tunnel work zone by sucking water out when needed. As the main structure rose, the gallery slabs were floated up to the next elevations, and the shafts were extended.

THE CONFLICT

The Queen's Chamber in the Great Pyramid may have been the original King's Chamber. The project boss was planning to use the levee slabs for fill after the original King's Chamber was finished. But the pharaoh had other plans, and that was when the Grand Gallery plan was born, with a new chamber for the king! This may be why the Grand Gallery seemed like an afterthought.

The Great Pyramid proved to be challenging and time consuming, and this was why they went back to simpler designs afterward.

EGYPTIAN LIGHT BULB EXPLAINED

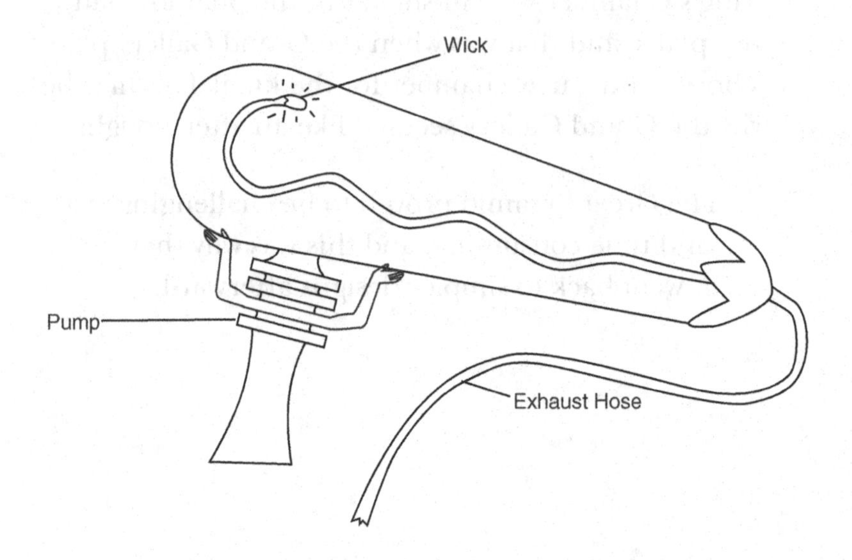

The Egyptians worked deep into cavern like situations. Some of these tunnels turned many corners and lighting became a challenge. Once again, they have an answer! Using the “hand and arm” powered pump to push fresh air into a glass chamber, enabled a wick to stay lit. Add on an exhaust hose and the toxic smoke is directed out of the work area. Note the modern looking nozzle attached to the glass vessel at the pump.

CONCLUSION

I believe that humans built the pyramids. It wasn't giants or aliens or magic wands. It was hard work with skill and planning that built them.

Dragging stones up a treacherous slope doesn't make sense. Some stones were shipped in from over 100 miles away. If you floated it that far, why not find a way to float it to its permanent location?

The Egyptians did it with style.

A LITTLE ABOUT THE AUTHOR AND WHY HE DECIDED TO ANSWER THIS QUESTION

Hello, my name is James A Kizer. I was born and raised in Northwest Indiana.
At the age of 11 my 5th grade teacher introduced the class to mechanical reasoning. She covered things like pulleys, levers and the incline plain. Then, went on to show us examples, like the ramp theories thought to have been used to build the Egyptian Pyramids. She also informed us that no one knew for sure, how they were built.

My little 11 year mind studied and calculated heights and distances for a few minutes and I came to the conclusion: The ramp idea didn't make sense and perhaps someday, I will figure it out!

So I went on with life. In high school, I usually tested above average in

areas like mechanical thinking and
spatial relations and before long,
was awarded an apprenticeship with
the Boilermakers, based on
testing well in these areas.

I served my apprenticeship and went on
working as a journeyman in
the trade. Sometime around my 15th year of boilermaking I
remembered that day in school, when I was 11.

After 15 years of heavy rigging, steel
construction, working with others,
planning and running work, I realized, I
now had the tools to answer the
illusive question: How were the Egyptian Pyramids built?

From time to time, as a boss, I would get
long periods of time to ponder
this question. Simply, by combining
my construction experience,
research, running models by trial and
error and adding in some prayer.

The theory slowly emerged. It took
nearly 12 years to develop. But,
once I got on the right trail, questions
started answering themselves.

Every theory I've ever heard, usually takes
me a short moment to poke
a hole, right through it. This theory holds water!
James A. Kizer

Printed in the United States
By Bookmasters